Jean Robert Pereira Rodrigues
Luana R. Pacheco Bittencourt

Study of the Machinability of Steels Based on Metallurgical Aspects

Jean Robert Pereira Rodrigues
Luana R. Pacheco Bittencourt

Study of the Machinability of Steels Based on Metallurgical Aspects

Steel Machinability Assessment

ScienciaScripts

This book is a translation from the original published under ISBN 978-613-9-66391-0.

Publisher:
Sciencia Scripts
is a trademark of
Dodo Books Indian Ocean Ltd. and OmniScriptum S.R.L publishing group

120 High Road, East Finchley, London, N2 9ED, United Kingdom
Str. Armeneasca 28/1, office 1, Chisinau MD-2012, Republic of Moldova, Europe
Printed at: see last page
ISBN: 978-620-8-01963-1

"But those who hope in the Lord shall renew their strength, they shall mount up with wings like eagles, they shall run and not be weary, they shall walk and not faint."

Isaiah 40:31 - Holy Bible

SUMMARY

Machining represents 50% of the production cost of mechanical components. The study of steels and their respective metallurgical aspects such as chemical composition, cold working, microstructure and inclusions have a marked effect on the machinability of this class of metallic material. The aim of this study is to assess the machinability of steel in relation to metallurgical aspects, by means of a survey of the literature. It was found that in cold working, in order to increase the hardness and decrease the ductility of steels, it is necessary to promote their hardening. It was also found that in terms of chemical composition and microstructure, it was possible to verify respectively that the phase change caused by heat treatment affects its machinability and that the increase in the sulphur content of the steel results in the formation of small chips during machining, favouring less flank wear of the cutting tool. It can be seen that knowing the machinability of steel, its characteristics and adding some alloying elements and knowing some process parameters can improve the process.

Keywords: Steels. Machinability. Metallurgical aspects. Process parameters.

SUMMARY

CHAPTER 1

INTRODUCTION

Materials have been of great importance in human culture since the earliest times for their use in various operations such as: transport, housing, communication, recreation, protection, work tools, etc. in short, in everything that is linked to their survival. However, the first human beings only had access to a limited number of natural materials (Passos, 2006).

Over time, scientists' understanding of the relationships between structural elements has enabled them to largely mould the characteristics of materials. As a result, tens of thousands of different materials have been developed with relatively specific characteristics that meet the needs of our modern, complex society; these include various types of materials such as metals, plastics, glass and fibres. The development of many of the technologies that make our daily lives so comfortable has been directly linked to the accessibility of suitable materials (Callister, 2006).

Machining is one of the most important processes in the production of mechanical components in the very broad world of manufacturing. It is estimated that around 10% of the gross domestic product of the United States is associated with machining processes, including tool costs and labour and capital investment costs (Shaw, 2005).

Steel has many fields of application in general, making it the most important metal alloy in the mechanical industry due to its good mechanical properties, the abundance of raw materials needed for its production and the fact that it is competitively priced compared to other metal alloys (PANNONI, 2009). The current market ranges from free-cutting steels with good machinability to tool steels that are difficult to machine. Among the main problems involved in machining steels is the production of long, continuous chips, a factor that makes machining this material difficult (MACHADO et al., 2009).

According to (PIMENTEL et al., 2015), steel machining in the production of mechanical components accounts for 50% of production costs. The machinability of a material is influenced by process parameters such as cutting speed, feed rate, cutting tool, lubrication, among others. However, the various types of steel show different results in relation to the type of machining used in the process. Therefore, the search for better machinability results and cost reduction leads to the production of different steels that are used to improve productivity rates in the manufacture of products through the machining process.

Steel materials are mainly composed of iron and carbon, often with a modest mixture of alloying elements. The biggest difference between cast iron and steel materials is the carbon content. Cast iron materials are compositions of iron and carbon, with a minimum of 1.7 per cent carbon to 4.5 per cent carbon. Steel has a carbon content of 0.05% to 1.5% (SCHNEIDER, 2009).

The aim of this work is to evaluate the metallurgical aspects of steels such as: microstructure, chemical composition, inclusions, cold working and process parameters. The aim is to get to know these characteristics in order to improve the machinability process.

CHAPTER 2

OBJECTIVES

2.1   GENERAL OBJECTIVE

Study of machinability based on the metallurgical aspects of steels, using the main machining processes as a basis.

2.2   SPECIFIC OBJECTIVES

- Analyse the effects of process parameters (Machining Force, Workpiece Surface Quality, Chip Formation; Cutting Temperature, Cutting Tool Wear) on machining;

- Discuss the factors that influence the machinability of steels, as well as studying the metallurgical aspects that interfere with machinability.

- To study cold working and evaluate the influence of rolled or studied materials on the machinability of steels;

- To study the effect of microstructure on the machinability of steels in the face of machining processes;

- To investigate the effect of the chemical composition of steels on their machinability;

- Search for techniques that make steel machining simpler and more economical;

CHAPTER 3

BACKGROUND

According to (TRENT, 2000) the machinability of a material is understood as the ease or difficulty of removing a material during machining and can be assessed through machining force (F_u), cutting temperature (T_c), surface finish, chip control, cutting tool wear rate; and other analyses, depending on the need, availability of infrastructure, facility, we can also take into account the mechanical vibration (v_b) of the workpiece-tool system, acoustic emission signals, among others.

Steels make up a large percentage of the materials used in the metal industry for various applications, due to their good mechanical properties, wide availability and relatively low cost. Although some steels are easy to machine, most of the metals in this group have poor machinability. Perhaps one of the biggest problems in machining steels is the production of continuous and usually long chips. Other factors also contribute to making machining difficult, including: the high melting point, the high temperatures developed on the surface of the tools (which happens at a certain distance from the cutting edge) and the high compressive stresses (SHAW, 2005).

Technological development has been aimed at obtaining materials that have good machinability. One of the important metallurgical factors in machinability is the presence of alloying elements. There are some alloying elements that have a positive effect on machinability, such as lead, sulphur and phosphorus, which are generally present in steels with improved machinability (DINIZ et al., 2006).

CHAPTER 4

LITERATURE REVIEW

4.1   GENERALISATIONS

According to (Amorim, 2003) machining is a term that covers mechanical manufacturing processes by generating surfaces through the removal of material, giving dimensions and shape to the part. A very broad definition of the term machining was described by Ferraresi (1977). According to him, machining is an operation that produces chips by giving the part its shape, dimensions or finish, or any combination of these three items.

In the process of machining a metal, at the start of the cut, the tool penetrates the workpiece material and it deforms elastically and plastically. This material will then exceed the maximum shear stress of the material and begin to flow. Depending on the geometry of the cutting wedge, the deformed material starts to form a chip that slides over the face of the cutting wedge. The material's performance in this machining process will characterise its machinability (BAPTISTA, 2002).

The machinability of a material can be understood as a response to the machining system. Considering that the most varied types of steel present different responses to a given machining system, it is difficult to study the machinability of this class of material. However, various techniques have been developed to facilitate steel machining and consequently reduce the costs involved in this process (PIMENTEL et al., 2006). These techniques range from precise control of the manufacturing/processing process, through the addition of chemical elements that promote chip embrittlement and/or lubrication of the cutting tool, to the engineering or modification of oxide inclusions resulting from the deoxidisation process (KLUJSZO, 2004).

As a general rule, pure carbon steel with a carbon content of less than 0.3% must be at its hardest possible state in order to achieve good machinability. This is achieved when it is cold drawn and has a fine grain. For high carbon contents (C > 7%) better machinability results

are obtained when the spheroidal structure is present. In alloy steels, alloying elements are generally added to increase hardness and produce a more resistant matrix, with a greater tendency to harden. From the point of view of machinability, alloy steels will produce better results than ordinary carbon steels (with the same carbon content) when their hardness is lower than that of ordinary steels (SHAW, 2005).

The addition of sulphur (along with manganese), lead or tellurium leads to the production of free-cutting steels (or easy-cutting steels). Steels deoxidised with calcium generate less wear on cutting tools than when deoxidised with silicon. The precise mechanism by which these additions improve machinability is not yet fully understood (possibly through the formation of a layer with constituents with lower shear strengths than the matrix itself), but they allow for higher cutting speeds, longer tool lives, better surface finishes, lower cutting forces, lower cutting temperatures, lower power consumption and better chip control (MACHADO et al., 2009).

The most commonly used tools for machining steels are high-speed steels and carbide. The limit of use for high-speed steel tools is the hardness of 300 HV of the steel. For carbide tools, this limit is higher, i.e. 500 HV (TRENT, 2000). When using high-speed steel tools, a crater develops due to diffusion and surface plastic shear. At higher cutting speeds, the tools can deform plastically under compressive stress. When using WC + CO carbide (K grade) at high cutting speeds, crater development and flank wear by diffusion will occur rapidly. Additions of TiC and TaC and NbC to tools (class P) will increase their life. At higher cutting speeds, edge deformation due to compressive stress can occur and the friction mechanism will prevail at low cutting speeds (TRENT, 2000).

4.2 A BRIEF HISTORY OF STEEL

The history of steel 1770 B.C. The first iron industry appeared in the south of the Caucasus, 1700 B.C., among the Hittites. The iron ore was in the form of small stones in the ground. The Hittites heated the mixture (ore and charcoal) in a hole in the ground and thus obtained a paste-like mass which was then beaten to remove the slag. What remained of the iron mass was then forged. The tool for producing iron was perfected and evolved into a semi-buried furnace in which layers of iron and charcoal were placed, according to the principles applied by the Hittites in their primitive furnaces (METAL TUBO, 2016).

Air blown through a manual bellows activated combustion. The temperature reached 1000-1200° C and a paste-like mass of iron weighing a few kilos was obtained by reduction, i.e. the elimination of oxygen from the ore. 1100 B.C. After the Caucasus, iron appeared in Egypt around 1100 B.C. Subsequently, it was found in regions that we now call Greece (1100 B.C), Austria (920 B.C), Italy (600 B.C), Spain France, Switzerland (500 B.C) (METAL TUBE, 2016).

According to (MALYNOWSKYJ et al., 2000) iron became even more important to humanity during the Industrial Revolution, which began in England at the start of the 19th century. However, the big change came with the discovery of steel in 1856, through the process created by Henry Bessemer, which allowed pig iron to be converted into steel by blowing oxygen, which led to a great improvement in the quality of the product, as it reduced the excess carbon (C), silicon (Si) and manganese (Mn) in the liquid pig iron. Later, this technique was associated with the Thomas process, created by Sydney Gilchrst Thomas, which eliminates excess phosphorus, improving the quality of steel and making it possible to expand the industrial process through the manufacture of steam engines and use in railway transport.

According to (IABR, 2016), steel is always present in our daily lives, such as Sunday lunch, on the way to work or in activities at home. Resistant, durable and 100% recyclable, steel is essential in the production of homes, vehicles, household utilities and consumer goods in general. It is so present in people's daily lives that it often becomes invisible. Steel is a fundamental material in construction in general. In a building or house, it makes its presence felt from the base to the finishes, providing security and stability for the entire structure.

4.2.1   Steel production

According to the Steel Industry in Brazil (IABR, 2016), steel mills around the world are classified according to their production process, as shown in figure 1:

>   **Integrated** - which operate the three basic phases: reduction, refining and rolling; they participate in the entire production process and produce steel;

>   **Semi-integrated** - which operate two phases: refining and rolling. These mills use pig iron, sponge iron or scrap metal purchased from third parties to turn it into steel in electric steelworks and then roll it. Furthermore, according to (IABR, 2016), the mills can also be categorised according to the products that predominate in their production lines:

- Semi-finished products (slabs, blocks and billets);

- Flat carbon steel (sheets and coils);

- Flat speciality/alloy steel (sheets and coils);

- Long carbon steel (bars, profiles, wire rod, rebar, wire);

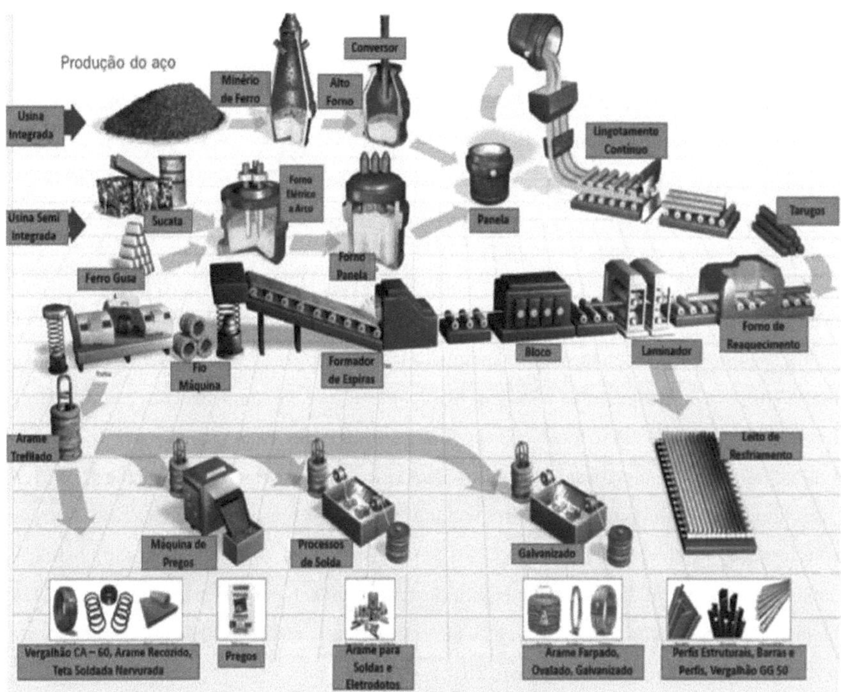

Figure 1 - Steel production
Source: GERDAU, 2016

(Gerdau Brasil, 2016) In integrated mills, there are two different reduction processes: the blast furnace, which produces pig iron, and the direct reduction, which produces sponge iron. In these units, steel is produced from iron ore found in nature in the form of rocks that need to be processed to obtain pig iron or sponge iron. In these plants, the transformation of the raw material into steel begins with the reduction process. In the blast furnace, the ore in granulated form is heated together with coke or charcoal to over 1400°C.

The steel is then transported to the continuous casting mill, where it is distributed in various veins, in cooling moulds to solidify into billets or blocks that will be cut to the appropriate size for rolling. Rolling mills produce slabs, bars and structural profiles, while wire rods are produced in the same way as steel.

annealed wire, welding wire and electrodes, barbed wire, galvanised wire and nails are produced through the wire drawing process (Gerdau Brasil, 2016).

Limestone, another type of flux, is added to help form the slag responsible for capturing the

11

ore's impurities. The result, after passing through the blast furnace, is iron in liquid form called pig iron. After this stage, the pig iron is transported to the converter in the steelworks where the metal is refined to transform it into steel. Steel is obtained by injecting oxygen into the liquid pig iron and also adding lime to promote the formation of slag. Depending on the desired end product, in the ladle the material may receive other alloying elements to give the steel new properties such as chromium and nickel, in the correct quantities in the case of austenitic stainless steels (Gerdau Brasil, 2016).

Sponge iron is obtained through an alternative process. Direct reduction takes place in a reactor. The process begins with the reduction of iron ore, which can be granulated or in the form of pellets. In the reactor, with carbon monoxide and hydrogen, the ore reacts at a temperature of 950°C and is transformed into sponge iron, which is highly metallised. In semi-integrated mills, steel production begins when the electric furnace is supplied with scrap, the main raw material, and inputs for the melting process. The metal is melted in the electric furnace and reaches temperatures of over 1700°C, resulting in steel in liquid form. After melting, the steel is transported to the ladle furnace. In the ladle furnace, the steel undergoes secondary refining to adjust its temperature and chemical composition. At this stage, the steel is alloyed (with chromium, nickel, molybdenum and others). This stage guarantees the production of quality steel, which is constantly evaluated through chemical tests based on established standards (Gerdau Brasil, 2016).

4.2.2 Stages

(IABR, 2016) Steel is basically produced from iron ore, coal and lime. Steelmaking can be divided into four stages: load preparation, reduction, refining and rolling, which can be seen in Figure 2 as the ore production flow.

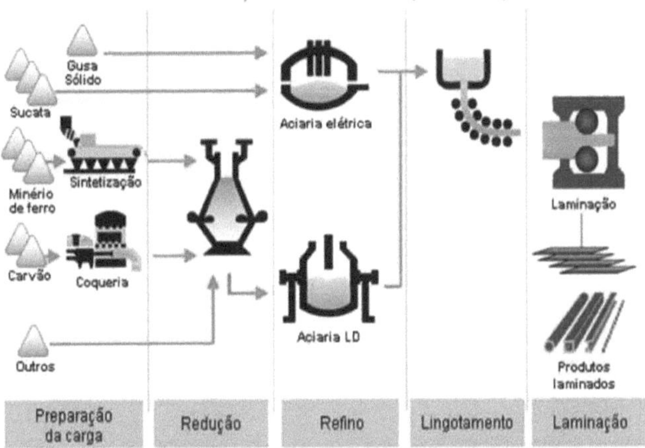

Figure 2 - Steel Production Flow.
Source: Brazil Steel Institute

4.2.2.1 Cargo preparation

-A large part of the iron ore (fines) is agglomerated using lime and coke fines;

-The resulting product is called sinter;

-Coal is processed in the coking plant and becomes coke.

4.2.2. 2Reduction

-These raw materials, now prepared, are loaded into the blast furnace;

-Oxygen heated to a temperature of 1000°C is blown out of the bottom of the blast furnace;

-The coal, in contact with oxygen, produces heat that melts the metallic charge and begins the process of reducing the iron ore into a liquid metal: pig iron;

- Pig iron is an alloy of iron and carbon with a very high carbon content.

4.2.2.1 Refining

- Oxygen or electric steelworks are used to transform liquid or solid pig iron and iron and steel scrap into liquid steel;

- In this stage, part of the carbon contained in the pig iron is removed along with impurities;

13

- Most liquid steel is solidified in continuous casting equipment to produce semi-finished products, ingots and blooms.

4.2.2.2   Lamination

The semi-finished products, ingots and blocks are processed by equipment called rolling mills and transformed into a wide variety of steel products, the nomenclature of which depends on their shape and/or chemical composition.

4.2.2.3   Flow

Semi-integrated mills operate two phases of the process: refining and rolling. Steel is obtained from the fusion of metals (scrap, pig iron and/or sponge iron) and refined in an electric furnace. Ferrous scrap is the main input (IABR, 2016).

4.2.3   The Importance of Steel and Its Main Applications

According to (PANNONI, 2009) steel is currently the most important of the metal alloys, as it is a fundamental material for the development of society. It has a strong presence in the daily lives of the human race, as it is present in most of the utensils used today, such as vehicles, buses, trains, civil construction, machinery, among others.

Because it has good properties, because it is a relatively low-cost material, because it has a large number of sources and suppliers worldwide and because it has an appreciable technical and scientific history of design and use, steel has a wide application in various industrial and commercial sectors. The main types of carbon steel and alloy steel classified by their application are shown below (CHIAVERINI, 2005):

- Casting steels: used for casting parts used in machine tools, the car industry, the railway industry and the shipbuilding industry.

- Structural steels: these are steels used mainly in construction and transport equipment: vehicles in general, road equipment, railways, ships, etc;

- Sheet steel: galvanised sheet; thin hot-rolled sheets, rolled bars, etc;

- Steel for wires, wires and springs: wires, flat wires; coil springs; nail making; piano wire, springs for lift cables and cranes, etc..;

- Easily machined steels: used to produce parts on automatic machines in the mass production process;

- Carburizing and nitriding steels: pins, small gears, rollers, camshafts; bearings;

- Pipe steels: pipes; structural pipes, pipes for oil pipelines;

    - Special-purpose steels: motors and transformers; high-temperature resistant; low-temperature resistant, etc...;

CHAPTER 5

INFLUENCE OF STEEL MACHINING ON PROCESS PARAMETERS

5.1   Machining force

(FERRARESI, 1977) states that knowledge of the machining force F or its components: cutting force F_c, feed force Ff and passive force F_p, is the basis as shown in figure 3:

* For the design of a machine tool (dimensioning of structures, drives, fixings, etc.);
* For determining cutting conditions under working conditions;
* To assess the precision of a machine tool under certain working conditions (deformation of the tool, machine and workpiece);
* To explain wear mechanisms.

Machining force is also a criterion for determining the machinability of a workpiece material. The components of the machining force (F_c, Ff and F_p) decrease with increasing cutting speed vc due to the decrease in material strength with increasing temperature (FERRARESI, 1977).

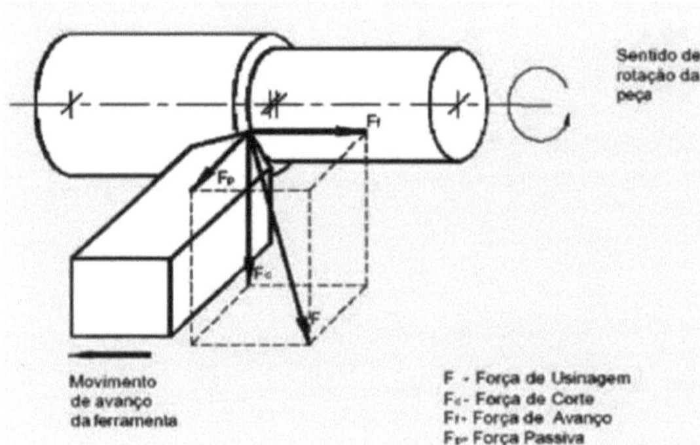

Figure 3 - Machining force components.
Source: MOLDES E INJEÇÕES, 2016

(DA SILVA, 2011) states that a piece of ABNT 1045 steel was dry machined, designed and manufactured in a suitable way for attachment to the dynamometer. The

The tests carried out simulated the cylindrical turning process. Tests were carried out to simulate orthogonal cutting at low cutting speeds.

(DA SILVA, 2011) found an increasing variation, observing linearity in the behaviour of the curve for material thicknesses in the range 1 p,m 2 20p,m, obtained when dry machining ABNT 1045 steel with cutting speed Vc = 150 mm/min, h = 1p.ni at 30pm, and a high speed steel tool, for thicknesses greater than 20pm the curve begins to vary between 237N and 614N as shown in figure 4.

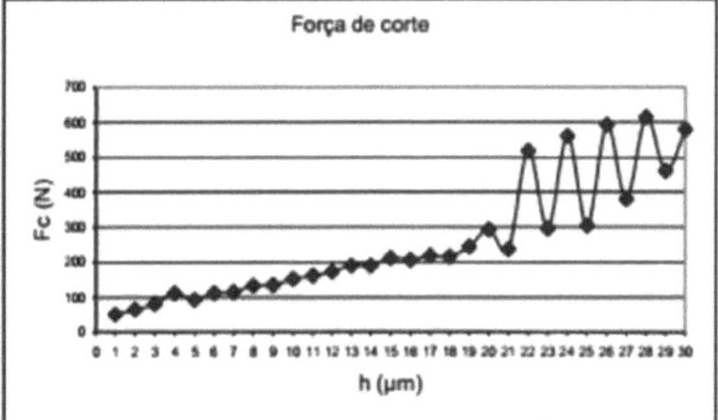

Figure 4 - Cutting force as a function of material thickness
Source: DA SILVA, 2011

Elastic deformation can be considered during cutting and this means that the tool needs to penetrate a minimum depth into the material before it starts cutting it and the force involved in this interval is called residual force, calculated by extrapolating a cutting force curve to a zero speed (DA SILVA, 2011).

5.2 Cutting Temperature

According to (FERRARESI, 1977) the influences of heat generation and transmission in metal machining are highly complex, as temperature rises change the physical and mechanical properties of the material being machined. Temperature, acting on tool wear, limits the application of higher cutting speeds and therefore sets the maximum conditions for productivity and tool life. Analysing the conditions of heat generation and transmission in the machining process as a function of different cutting factors allows for the choice of more convenient tool dimensions and shapes, as well as improving the working regime and durability.

A widely used method for roughly estimating the temperature field during machining is the numerical method. The resolution is based only on input data, and the machining operation does not necessarily have to be carried out. Due to recent advances in computing, this technique has become very widespread. The most common methods found in the literature for solving direct thermal problems are: finite elements, finite differences and finite volumes (BARRIOS, 2013).

(BARRIOS, 2013) estimated the temperature field in the chip, tool and workpiece during the milling of AISI 1050 and H13 steels, using uncoated carbide tools. A numerical model based on the finite difference method was used. Figure 5 shows the chip temperature distribution for a tool inclination angle of 6° and a cutting speed of 80 m/min.

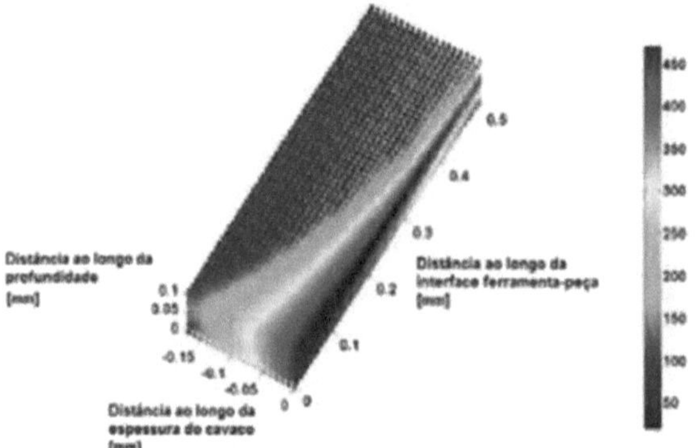

Figure 5 - Chip temperature distribution.
Source: BARRIOS, 2013

This method generally uses devices or sensors to quantify the temperature gradient using electrical signals. The

factors that can be considered when choosing a method are: temperature range, sensor robustness, temperature disturbance by the sensor, signal quality in terms of noise, response time and uncertainties (BARRIOS, 2013).

5.3   Cutting Tool Wear

Tool life is very important for machining processes, especially when it comes to the mass production sector. In the studies of (CASAGRANDA, 2004), tool life can be described as the time that the tool works effectively until it loses its cutting capacity, according to previously established parameters. Tool life is also a criterion observed when determining the machinability of a material, because after the determined limit the machining of the part will be subject to dimension errors and poor surface quality.

The classification of wear types is an important aid for estimating and optimising machining by correctly specifying inserts and machining data for a given type of machining and material. Cutting tool life can be controlled and increased by inspecting the extended cutting edge and acting on the type of wear. In order to have the right insert, there are important factors for achieving optimum wear development such as: good initial cutting data values, expert support, experience, good workpiece material quality and machines in good condition (NASCIMENTO, 2008).

When machining with a high speed steel tool, the destruction of the cutting edge is due to the decrease in hardness caused by the increase in cutting temperature. This happens at temperatures of around 600°C. This can be observed in two ways: by the formation of a shiny ring on the cutting surface of the workpiece, usually accompanied by a loud noise, and by the increase in the feed and depth components of the machining force. In finishing operations, in general, the same considerations apply as for carbide. However, the cutting speeds when using high speed steel are relatively low, jeopardising the finish of the workpiece. Wear, limited by the tolerance and roughness requirements of the workpieces, results in a very short tool life (DINIZ et al., 2006).

One factor influencing tool wear is machining conditions, which means that the progression of wear is influenced by the cutting speed, then by the feed rate and finally by the machining depth. So, for example, the decrease in tool life caused by a 10 per cent increase in cutting speed is much greater than that which would occur if the feed rate were increased by the same proportion. As such, cutting speed is the parameter that most influences wear, because increasing it increases the energy (heat) that is imputed to the process, without increasing the area of the tool that receives this heat (DINIZ et al., 2006). Increasing the feed rate also increases the amount of heat imparted to the process, but also increases the area of the tool that receives this heat. Table 1 shows some tool life data for various feed rates and cutting speeds.

($_{ap}$ = 1mm, Workpiece material: 4340 steel, Coated carbide tool class P35)

Table 1 - Turning Tool Life (Cutting Path) for Different Feed Rates and Cutting Speeds

f (mm/turn)	Vc (m/min)	Life (m)
0.16	300	1450
0.20	300	1530
0.24	300	1550
0.20	250	2600
0.20	300	1530
0.20	350	650

Source: DINIZ et al., 2006

5.   4Surface finishing   of the part

The quality of the surfaces obtained during machining can be a criterion for determining the input parameters. The factors influencing surface quality are listed in figure 6 (WEINGAERTNER, 2003).

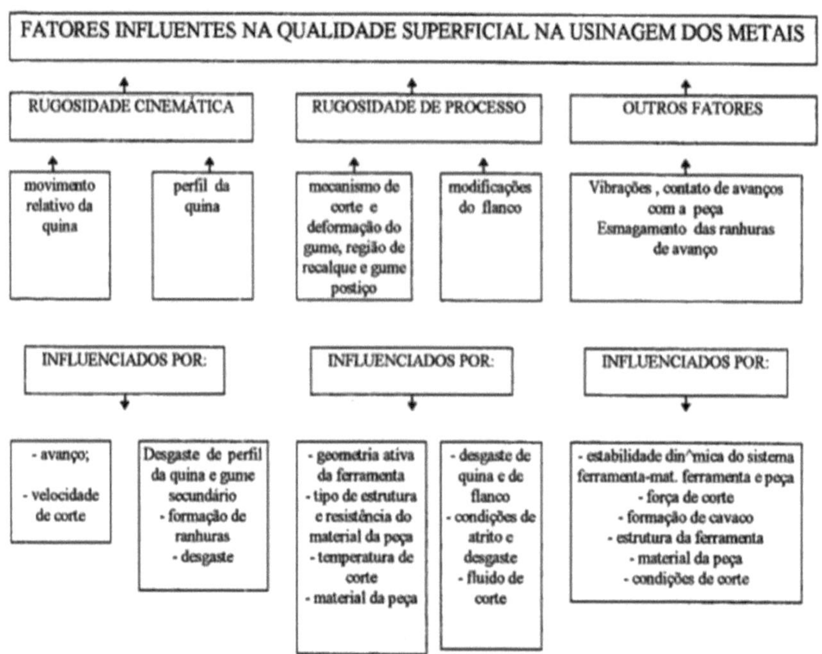

Figure 6 - Factors influencing surface quality.
Source: WEINGAERTNER, 2003.

(AKASAWA et al., 2003) when machining resulphurised steels with and without alloying elements to improve machinability (Bi, S, Cu and Ca), observed that the presence of manganese sulphides led to greater roughness, i.e. worse surface finish on the workpiece. (JIANG et al., 1996) also carried out tests on the same type of steel and observed that the APCs also influence surface quality, as smaller cutting edges provided a better surface finish, concluding that the size of the APC influences the surface finish.

When turning hardened steels, it is difficult for a false edge to form due to the low ductility of the workpiece material and the high cutting temperature. As a result, the cutting edge is transferred to the machined surface with reasonable precision. So, as long as the wear on the gap surface remains small and uniform, the roughness will not increase. As wear progresses, however, the cutting edge tends to "thicken" and consequently deteriorates the machined surface. Average roughness values in steel turning are in the range of 0.1 to 0.4 um, depending on the cutting parameters and workpiece material (COSTA, 1993).

5.5 Cavaco's formation

According to (DINNIZ et al., 2006), the chips formed in machining processes can be classified according to three types: continuous, shear and rupture. The first is characterised by juxtaposed lamellae in a continuous, grouped arrangement. The second type has the same configuration, but the lamellae are more defined and partially welded along the chip. The main characteristic of the third type is the complete rupture of the lamellar segments. Chips can also be differentiated in terms of their shape, which can be ribbon, helical, spiral and chip. These shapes are shown in figure 7.

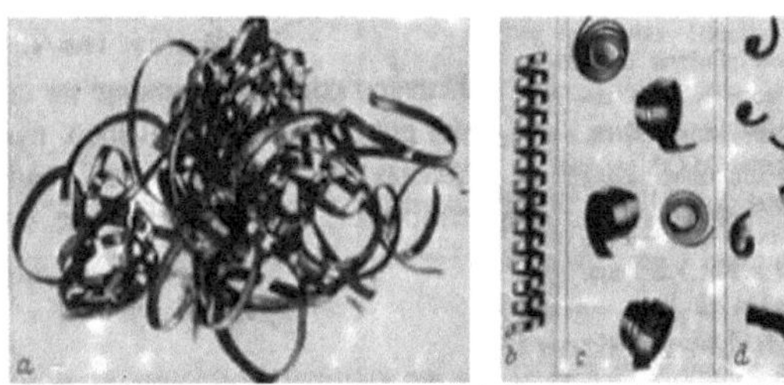

Figure 7 - Chip shapes: (a) ribbon, (b) helical, (c) spiral, (d) chip.
Source: DE SOUZA , et al., 2006

(DE SOUZA , et al., 2006) and their colleagues claim that chips can be a very important element of machining research, even though they don't seem to be and the vast majority of professionals who deal with manufacturing, especially in companies, dismiss this fact. As a rule in industry, chips only become the main focus when they interfere negatively with the end product. It is true that the main result to be achieved is the machined product and not the material removed from it. However, studying chips can provide important information for understanding the process and, consequently, optimising it.

According to (DE SOUZA , et al., 2006) they carried out different turning tests on ABNT 1045 steel, with an initial diameter of 35.8 and a cutting length of 300 mm, using dry carbide tools. The cutting parameters used were feed rate f = 0.094 and 0.19 mm/rev, machining depth íE, - 0.5 and 1.0 mm and cutting speed E -86.117 and 134m/min. The tests for different cutting parameters showed a variety of chip shapes, types and colours, as shown in Table 2.

Table 2 - Summary of input data and investigation of the chips generated.

Essay	Parameters		north	Cavaco classification		
	- (m/me)	rc , [mm]	f[mm/rev]	Type	Shape	Colour
1		0,5	0,094	Continuous	Helical	Yellow
2			0,190	Continuous	Helical	Yellow
3	86	1,0	0,190	Continuous / Rupture	Helical / Chips	Blue
4			0,094	Rupture	Chips	Blue and copper
5		0,5	0,094	Continuous .	Ribbon / Helical	Dark blue and copper
6	134-		0190	Continuous	Helical	Blue and copper
7		1,0	0,190	Continuous	Helical	Blue and

							copper
8				0,094	Continuous .	Ribbon Helical	Light Blue
9	.	117	0,5	0,297	Continuous	Helical	Blue
10				0,449	Continuous	Helical	Yellow

Source: DE SOUZA , et al., 2006

(DE SOUZA , et al., 2006) and their collaborators found in test 1 that the chip formed was of the continuous and helical type. The low feed rates and machining depths meant that the chips removed were long (0.75 to 1.00 metres). In test 2, shown in figure 8, it can be seen that the increase in feed caused the chip to break into smaller fragments than in the first test (20 to 50mm). The slightly yellowish colour of the chips in both tests shows that the heat generated in the cutting zone was small, preventing further oxidation of the removed chip.

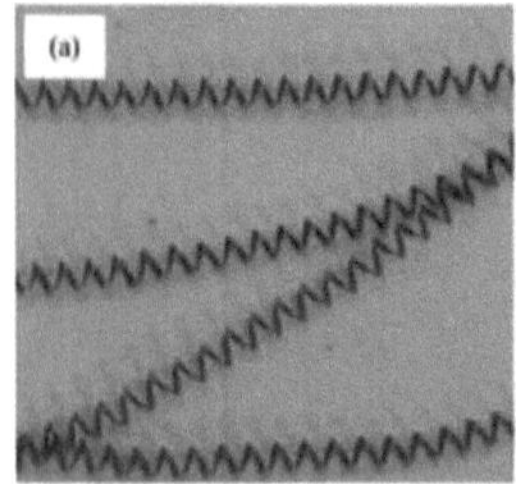
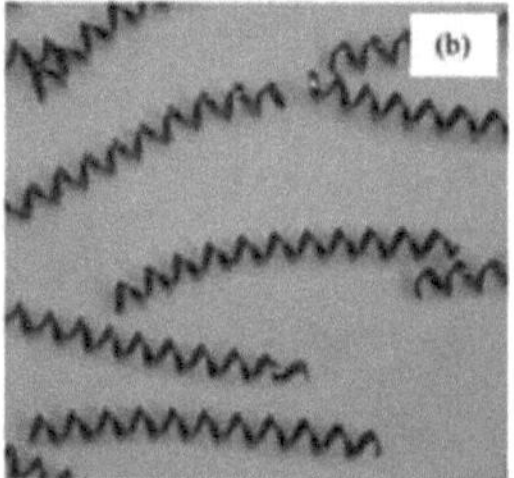

Figure 8 - Chips obtained in tests 1 (a) and 2 (b).
Source: DE SOUZA , et al., 2006

(DE SOUZA , et al., 2006) and their colleagues observed the presence of continuous and breaking chips in the 3rd test. The increase in machining depth caused a large increase in the temperature of the chip-to-tool contact zone, characterised by the blue colour of the chip, which is peculiar to cases where there are high cutting temperatures. However, (DE SOUZA , et al., 2006) noted the presence of rupture chips, which are characteristic of machining conditions where there are higher shear rates. They state that increasing the feed rate by two times may have favoured the formation of a shear chip. In the 4th test (DE SOUZA , et al., 2006) and other researchers found a bluish colour shown in figure 9, which confirms that the increase in machining depth promoted an increase in temperature in the cutting region, however the presence of chips allows us to estimate that the contact between the removed chip and the exit surface was reduced.

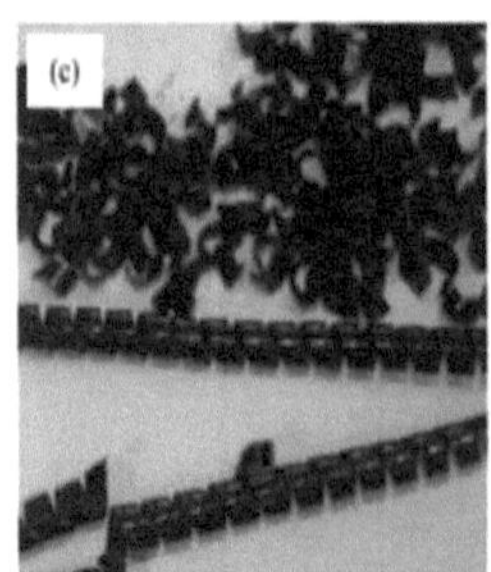

Figure 9 - Chips obtained in tests 3 (c) and 4 (d).
Source: DE SOUZA , et al., 2006

(DE SOUZA , et al., 2006) when evaluating the 5th test, as shown in figure 10, realised that with the 56% increase in cutting speed, there was the formation of continuous ribbon and helical chips. Compared to the first test, where only the cutting speed was different, they noticed a significant change in the shape of the chips generated. They noticed that the bluish colour indicates greater heat generation between the workpiece and the tool, when compared to the increase in cutting speed.

(DE SOUZA , et al., 2006) and their colleagues, when carrying out test 6, found that the chip obtained was very similar to the one obtained in test 2, except for the bluish colour. Once again, they explain that this is due to the fact that the increase in cutting speed promotes an increase in the shear rate of the material and consequently raises the temperature in the cutting zone (DE SOUZA , et al., 2006).

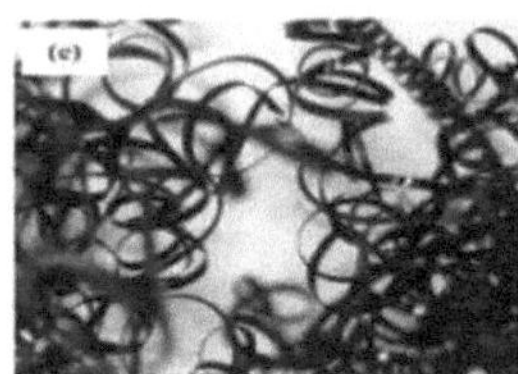
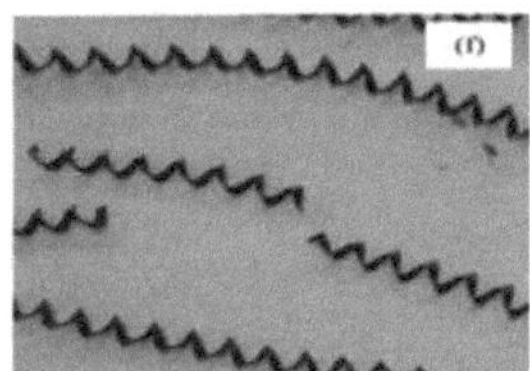

Figure 10 - Chips obtained in tests 5 (e) and 6 (f).
Source: DE SOUZA , et al., 2006

(DE SOUZA , et al., 2006) and their collaborators found that in test 7 (figure 11), there was a significant reduction in the size of the continuous helical chip when compared to that obtained in test 3 (from 40 to 20mm). However, no chipped chips were found in test 7. On the other hand, in test 8, at different times, two types of chip were observed. At first, the chip formed was classified as helical and long. This was followed by the formation of a tangle of ribbon chips that intertwined with the test specimen. When the ribbon chip automatically broke, a continuous helical chip was formed, which then broke and was replaced by the ribbon chip. The light blue colour indicates that there was considerable heat generation, but less than in the previous 7° test (DE SOUZA , et al., 2006).

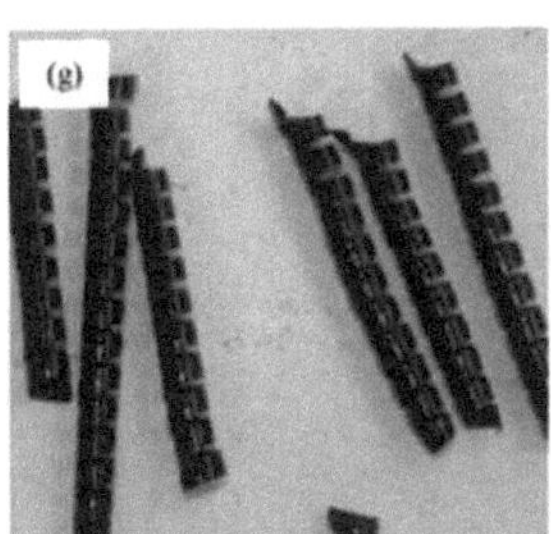
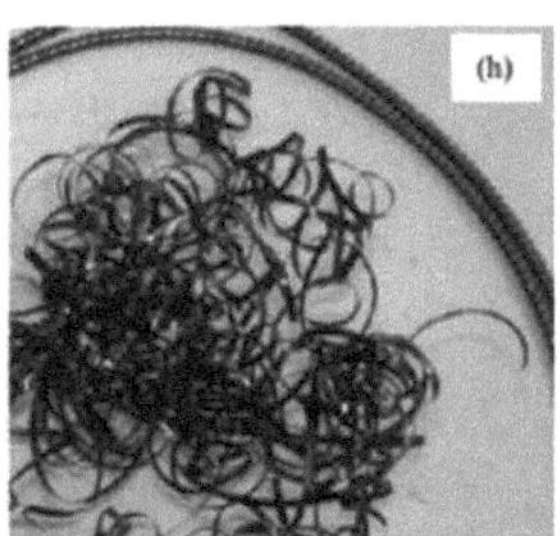

Figure 11 - Chips obtained in tests 7 (g) and 8 (h).

For trials 9 and 10, as shown in figure 12 (DE SOUZA , et al., 2006) and their collaborators confirmed that they had the same characteristics as the previous ones, except for the use of very high feed rates. The size of the chips in these two tests differed very little, but the big difference was in their colouring. (DE SOUZA , et al., 2006) and his colleagues noticed in the 10th test that the colouring was quite yellowish, while in test 9, the hue was bluish.

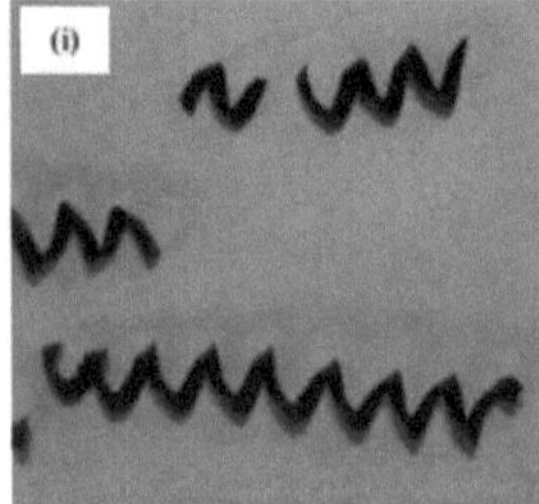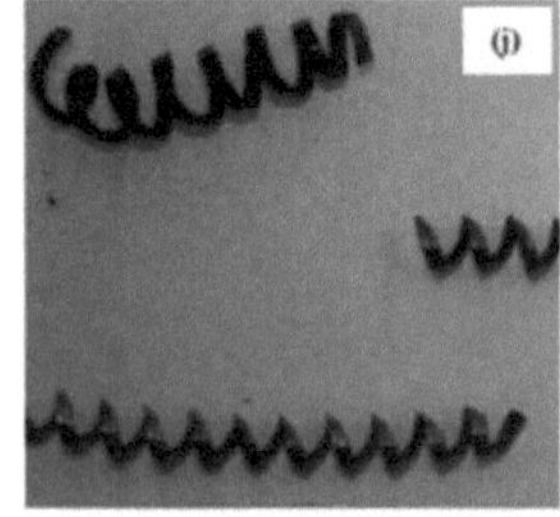

Figure 12 - Chips obtained in tests 9 (i) and 10 (j).
Source: DE SOUZA , et al., 2006

Analysing the results obtained, (DE SOUZA et al., 2006) state that all the cutting parameters used as input variables in the experiment had an influence on chip formation. In general, the results were in good agreement with those presented in the literature. The increase in cutting speed meant that the portion of the workpiece material removed in the form of a chip suffered higher shear rates, in some cases changing the shape of the chip from helical to ribbon. In addition, the most pronounced effect relates to chip colour, indicating that higher shear rates generate more temperature and heat at the chip-to-tool interface. The type of chip remained typically continuous, with the exception of tests 3 and 4 conducted at low cutting speeds and greater machining depths (DE SOUZA , et al., 2006).

(DE SOUZA , et al., 2006) found that the increase in machining depth may be or may be associated with a change in chip shape from chips where Vc is lower to strips where Vc is higher. Increasing the feed rate led to the formation of helical chips, unlike in test 3. To a certain extent, this result can be confirmed by analysing the results of tests 9 and 10, in which large feed rates were used.

CHAPTER 6

METALLURGICAL ASPECTS THAT AFFECT THE MACHINABILITY OF STEELS

The predominant metallurgical factor with regard to machinability is obviously hardness. Low carbon steels with low hardness and high ductility have a tendency to form a cutting edge, resulting in reduced tool life and deterioration of the surface finish. Adding a higher percentage of carbon improves machinability due to the increase in hardness and decrease in ductility. In terms of the influence of steel hardness on machinability, it can be said that 200 HB is an average value. As the hardness decreases below this value, the tendency for the cutting edge to form increases. When the hardness increases above this value, tool wear via abrasion and diffusion becomes a factor that negatively affects the machinability of the material (DINIZ et al., 2006).

The other metallurgical factor of importance in machinability is the presence of alloying elements. Some alloying elements have a positive effect on machinability, such as lead, sulphur and phosphorus, which are generally present in steels with improved machinability. On the other hand, carbide-forming elements (which are hard, abrasive particles) such as vanadium, molybdenum, niobium and tungsten, as well as others such as manganese, nickel, cobalt and chromium, have a negative effect on machinability. A study of a number of metallurgical aspects is shown below: (DINIZ et al., 2006).

(CHIAVERINI, 2005) comments that high hardness values mean machining difficulties, while medium and low values are associated with good machinability properties, but that hardness measurements would not serve as an absolute reference for determining true machinability.

6.1   Cold working

The cold deformation process can improve machining through the hardening it causes, which increases hardness and makes it more difficult for the chip to stick to the tool and form the cutting edge. If the hardness of the workpiece is excessively high, there will be excessive tool wear and an increase in the power required (BAPTISTA, 2002).

The rate of tool wear is reduced as a result of cold working, due to the reduction in cutting energy required and the drop in working temperature. In materials with a high rate of hardening and high ductility, the cold working process causes a sharp drop in machinability (GONZALES, 1992).

A good measure to increase the hardness and decrease the ductility of low carbon steels (which normally have a hardness well below 200 HB) is to promote their hardening via cold working. Figure

13 shows the comparison in terms of tool life for a 1016 steel (low carbon) in several different machining operations. It can be seen in this figure that tool life increased in all cases after the bars of this steel were cold drawn and then machined, an operation which caused the hardness of the parts to increase from around 125 HB to 180 HB (DINIZ et al., 2006).

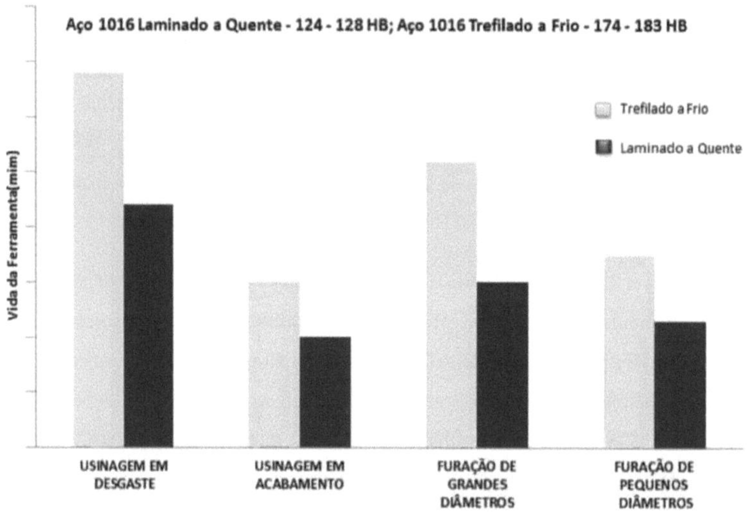

Figure 13 - Effect of hardening (by cold drawing) on tool life.
Source: DINIZ, 2006

6.   2Microstructure

(DINIZ et al., 2006) and their collaborators point out that a second metallurgical factor to consider that does affect machinability is microstructure. Figures 14 and 15 below show how microstructure variation, via phase change caused by heat treatment, affects machinability. As can be seen in figure 14, the martensitic structure is very hard and resistant and generates very low carbide tool life.

(DINIZ et al., 2006) and their colleagues state that steels with abrasive structures can only be machined efficiently using abrasive processes or ultra-resistant tools. As can be seen in Figure 15, when you go from an alloy with 10% ferrite and 90% pearlite to an alloy with 35% ferrite and 65% pearlite, the tool life increases considerably, even though the hardness of the workpiece has decreased by around 6%. He says that the reduction in micro-constituents such as pearlite and cementite reduces the amount of abrasives, which are responsible for wear in tool life. This is due to the fact that when the pearlite content decreases, so does the cementite content (pearlite is a combination of ferrite and cementite).

26

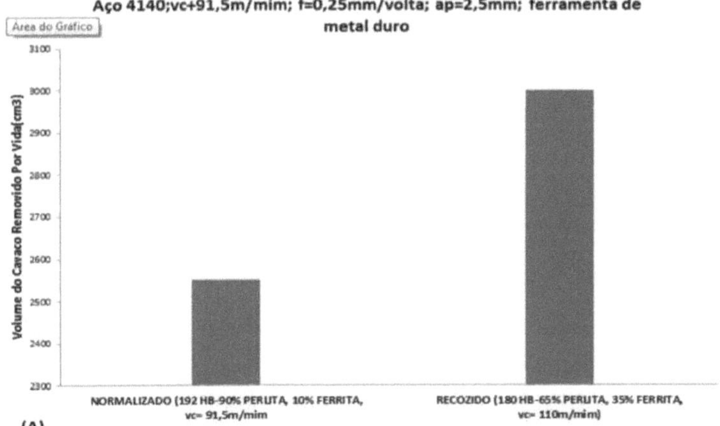

Figure 14 - Effect of Microstructure on the Machinability of 4140 Steel.
Source: CHIAVERINI, 2005

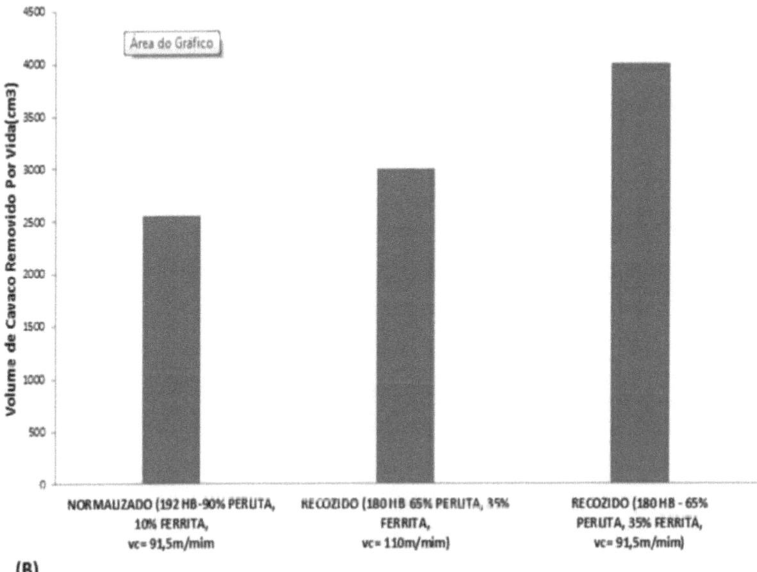

Figure 15 - Effect of Microstructure on the Machinability of 4140 Steel.

4140 steel, f=0.25 mm/turn, ap=2.5mm, carbide tool
Source: CHIAVERINI, 2005

The effect of microstructure on the machinability of steels can be summarised as follows: a) for very low carbon steels (up to 0.20%), the most recommended condition, even for reasons of economy, is to simply roll them, but better results are obtained if stress relief is applied. The purpose of this stress relief is to remove the hardening left over from rolling (DINIZ et al., 2006).

According to Chiaverini (apud Gilberto Sanches Gonzales, 1992, p.13) we can evaluate the effect of the microstructure of steels based on their carbon content, as shown in table 3.

Table 3 - Best Machining Conditions Depending on the Carbon Content of the Steels

CARBON CONTENT	BEST CONDITION	HEAT TREATMENT
(a) Up to 0.2%	Pearlite b.	laminate (more economical)
(b) 0.2 to 0.3%	Pearlite b. laminated	Norm. Improvement
(c) 0.3 to 0.4%	Coarse pearlite	Annealing improves
(d) 0.4 to 0.6%	Pearlite or Spheroidised	Full annealing
(e) Higher 0.6%	Spheroidised	Coalescence

Source: GONZALES, 1992

(FERRARESI, 1977) points out that because it is possible to change the microstructure of steel or cast iron without altering the chemical composition, this is an important factor influencing machinability. The microconstituents alter the ductility and brittleness characteristics depending on their presence, quantity and shape, promoting different chip breaking conditions, abrasiveness, cutting force and temperature. The presence of acicular phases such as bainite and martensite, due to their extremely abrasive effect, is also undesirable in machining (AMARO, 1982).

6.3 Chemical composition

6.3.1 Effect of Sulphur

(PIMENTEL, 2006) states that good machinability in low-carbon steels is achieved by increasing the sulphur content to values above 0.40%. He states that because it is not an expensive additive, sulphur is the element most widely used to improve the machinability of carbon steels. The author points out that in almost all types of steel produced commercially, sulphur is generally combined with manganese in the form of manganese sulphide inclusions. (PIMENTEL, 2006) concludes that the characteristics of these types of inclusions, such as size, morphology, frequency and distribution, have a marked influence on the machinability of free-cutting steels.

In order to understand the influence of sulphur on machinability, (PIMENTEL, 2006) carried out turning tests on bars with a diameter of 22.23 mm containing the following composition: 0.14% carbon, 0.80% manganese and explains that the sulphur content varied between 0.025 and 0.25%. After the tests (PIMENTEL, 2006), he found, as shown in figure 16, that for a given feed rate, increasing the sulphur content reduces the deformability of the chip, i.e. the amount of plastic deformation associated with chip formation.

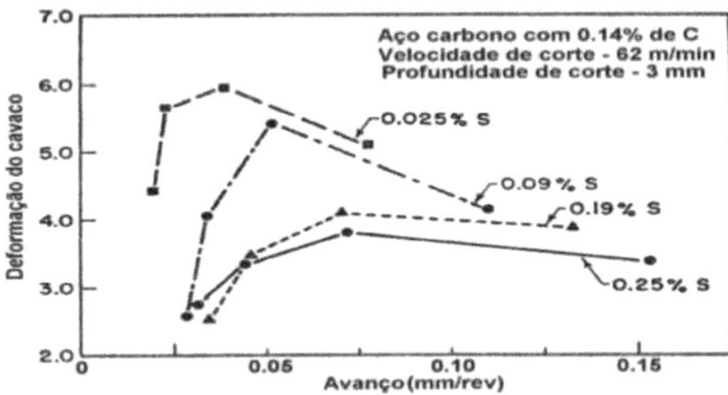

Figure 16 - Relationship between Sulphur Content and Chip Deformation at Various Feed Rates in Turning Tests.

Source: PIMENTEL, 2006

(PIMENTEL, 2006) states that the effect of sulphur on chip size is shown schematically in Figure 17. (PIMENTEL, 2006) noted that chip size decreases with increasing sulphur content, providing a smaller contact area between the chip and the cutting tool. Consequently, the friction forces between the chip and the cutting tool are lower, causing less intense wear of the cutting tool.

Figure 17 - Relationship between sulphur content and chip size.

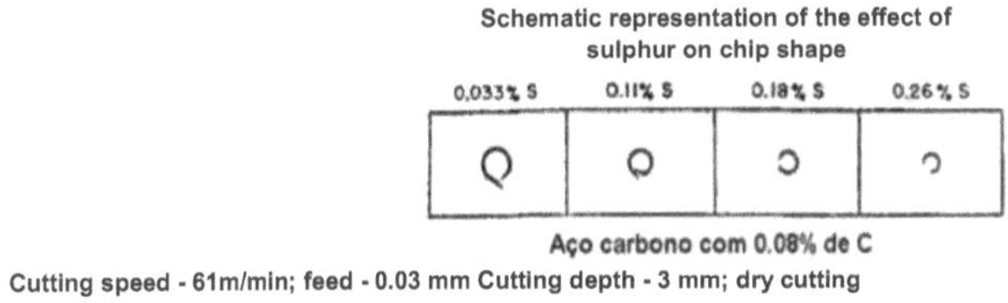

Source: PIMENTEL, 2006

(PIMENTEL, 2006) concludes that increasing the sulphur content in steel results in the formation of small chips during machining. He states that smaller chips mean a reduction in the contact area between the chip and the cutting tool. Consequently, friction at the chip-cutting tool interface is lower, leading to an increase in cutting tool life. In addition, (PIMENTEL, 2006) found that due to the presence of manganese sulphide inclusions, friction at the tool-workpiece interface is also lower, favouring less cutting tool flank wear (PIMENTEL, 2006).

6.3.2 Effect of Carbon and Silicon

(PIMENTEL, 2006) points out that although the international AISI and SAE standards allow carbon contents above 0.13% in resulphurised and rephosphorised low-carbon steels, most of the easy-cutting steels frequently used have carbon contents below 0.10%. He states that this fact can be seen by carefully analysing the curves in Figure 18, where it can be seen that the highest machinability indices are obtained for low carbon contents (below 0.10%). In addition, these curves also show the

29

existence of a **carbon content** ("carbon peak"), above which the machinability index decreases (PIMENTEL, 2006).

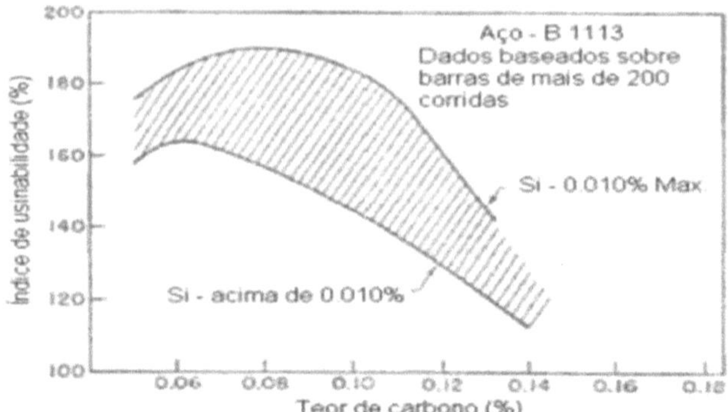

Figure 18 - Effect of Carbon Content on the Machinability Index of B1113 Steel Bars in Turning Tests.
Source: PIMENTEL, 2006

In Figure 18, he also observed that silicon has a detrimental effect on the machinability of low-carbon steels. This effect is shown more clearly in Figure 19, where he found that the machinability index decreases linearly as the silicon content increases from 0.003% to 0.03% (PIMENTEL, 2006).

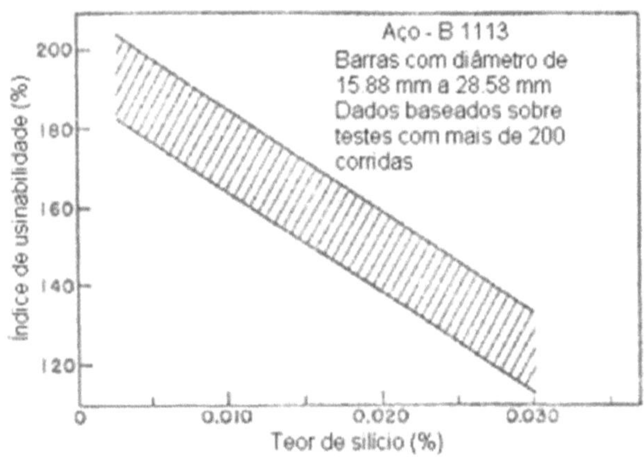

Figure 19 - Effect of Silicon Content on the Machinability Index of B1113 Steel Bars in Turning Tests.
Source: PIMENTEL, 2006

(PIMENTEL, 2006) found that the use of silicon at around 0.20% increases the layer of manganese sulphide deposited on cutting tools when machining free-cutting steels, thus improving the steel's machinability. He also discusses the use of silicon as a way of improving machinability.

6.3.3 Effect of Phosphorus and Nitrogen

(PIMENTEL, 2006) states that within certain limits, phosphorus benefits the machinability of free-cutting steels by improving the surface quality of the machined part and by contributing to the formation of brittle chips. Moderate additions of nitrogen also act in a similar way.

(PIMENTEL, 2006) found that while on the one hand small additions of nitrogen and phosphorus improve the machinability of free-cutting steels, on the other hand large additions of these elements can be highly damaging due to the reduction in cutting tool life. For example, the author notes, as can be seen in Table 4, that increasing the nitrogen content from 0.003% to 0.028% has a detrimental effect on the tool life obtained when machining free-cutting steels with low and high phosphorus contents. In addition, he also notes that for high nitrogen levels the best machinability performance is exhibited by steel with a residual amount of phosphorus (0.018%) compared to rephosphorised steel (0.12%). Therefore (PIMENTEL, 2006) states that controlling the levels of these elements is necessary to guarantee an optimum combination of tool life, surface finish of the machined part and cold forming capacity.

Table 4 - Effect of Nitrogen Content on the Performance of Drawn Steel Bars Used in the Production of Screw Nut Drums.

Steel	P.%	N_2 , %	Average tool life, hours			
			77 m/min	86 m/min	95 m/min	105 m/min
1	0,018	0.003	b	b	6.2	7.3
1	0.018	0.028	b	12.1	3.3	b
2	0.09	0.003	b	b	b	8.4
2	0.12	0.028	12.0	3.8	1.6	1'
a	Chemical	Steel 1	0.09% C	0.87% Mn	0.29% S	0.008% Si
	composition	Steel 2	0.09% C	0.96% Mn	0.28% S	0.004% Si
b	Not tested					
Note	All tests were carried out on drawn bars with a diameter of 19 mm					

Source: PIMENTEL, 2006

6.   4Inclusions

(PIMENTEL, 2006) states that all commercially produced steels contain a measurable quantity of inclusions, which can be endogenous, i.e. intrinsic to the manufacturing process and the result of chemical reactions occurring within the metal, or exogenous, i.e. of external origin and the result of mechanical erosion of the refractory or the chemical reaction between the metal and the refractory or between the metal and the slag.(PIMENTEL, 2006) With regard to endogenous inclusions, certain types, when properly controlled, can become desirable, and this is what happens in steels where machinability is the main concern. Table 5 shows the inclusions that most affect the machinability of steels.

Table 5 - Inclusions present in steels.

Category	Effect	Example	Form of Presence in Steel

31

Metallic:			Up to 0.24% vol. in lead steels. Present as elongated particles in hot-rolled products or
-Pb/ -Bi	Positive	Pb	in sulphide tails.
Hard non-metals:			Less than 0.05% vol. in deoxidised steels with aluminium .
-Aluminates/ -Nitrides	Negative	$Al\ O_{23}$	Present singly or in streaks, with particles < 5 µm. They can be modified with calcium.

Source: PIMENTEL, 2006

In the case of metallic inclusions, which have a positive effect, these inclusions are present as elongated particles in laminated products. Hard non-metallic inclusions have a negative effect and are present in isolation or in striations (PIMENTEL, 2006).

6.4.1 Effect of Metallic Inclusions

(PIMENTEL, 2006) states that although lead or bismuth can theoretically be added to any type of steel, these metallic additions are usually limited to the free-cutting steels of the SAE 11XX and 12XX series, as the combination of lead and manganese sulphide has been shown to provide the greatest machinability in steels. He states that steels with added lead are identified by the inclusion of the letter L between the second and third digit

in the SAE/AISI classification, such as SAE 12L14 or SAE 10L45. The lead content varies between 0.15 and 0.35%, and its addition alone does not cause a significant reduction in mechanical properties at room temperature. At temperatures close to its melting point, lead can cause the steel to embrittle. As an example, the effect of lead on the machinability of 1213 steel can be seen in Figure 20. From this graph, considering a certain percentage of equivalent phosphorus, it can be seen that lead improves the productivity of 1213 steel by approximately 40% (PIMENTEL, 2006).

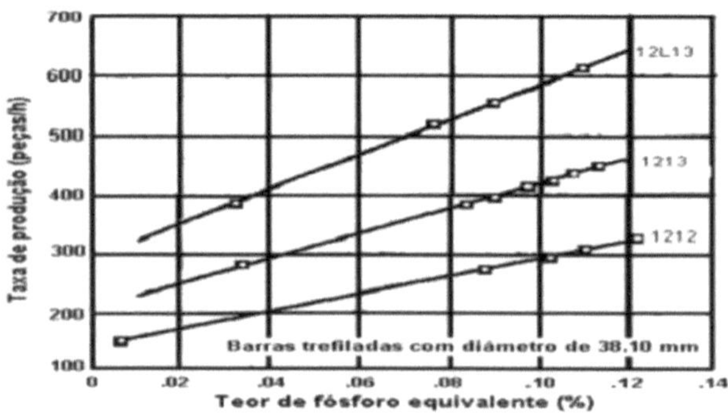

Figure 20 - Effect of the Equivalent Phosphorus Percentage on the Production Rate for Grade 12XX Steels, with and without the addition of Lead.

Source: PIMENTEL, 2006

(PIMENTEL, 2006) comments that for an equivalent phosphorus percentage of 0.09%, the addition of lead provides an increase in productivity from 400 pieces/h to 560 pieces/h. This 40% increase in productivity is due to the cutting speeds and tool feeds that can be increased, as can be seen in Figures 21 and 22. And he mentions that in the same way as with grade 12XX steels without added lead, an increase in the equivalent percentage of phosphorus improves the machinability of lead steels, due to the possibility of operating under higher cutting conditions without sacrificing surface quality or compromising the life of the cutting tool (PIMENTEL, 2006).

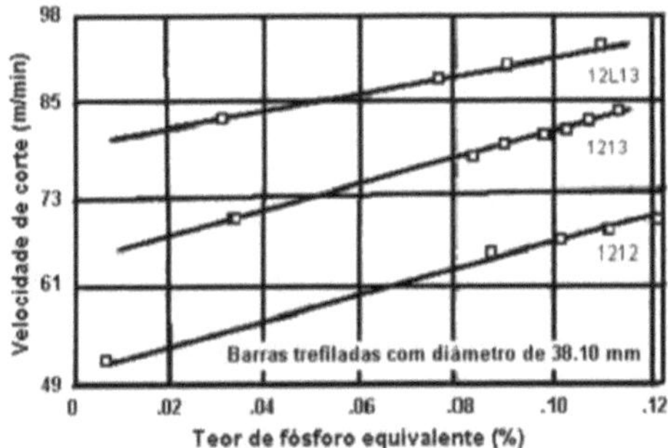

Figure 21 - Effect of the Equivalent Phosphorus Percentage on Machining Conditions for Grade 12XX Steels with and without Lead Addition.

Source: PIMENTEL, 2006

Figure 22 - Effect of the Equivalent Phosphorus Percentage on Machining Conditions for Grade 12XX Steels with and without Lead Addition.

Source: PIMENTEL, 2006

Due to environmental and health concerns during the steelmaking process, the use of lead as an added element has been greatly restricted, to the point where certain countries have abandoned its production. However, due to its superior machinability, components made from leaded steel are still used by practically all car manufacturers. Despite being on the market for around 30 years, bismuth steels, created to replace lead steels, still have inferior performance, especially at low and medium cutting speeds, which restricts their use as a possible replacement. According to (PIMENTEL, 2006), for the same machinability result, only a third of the steel is machinable.

part of bismuth compared to lead should be used. This lower addition translates into greater fatigue resistance of the components made from the steel, as well as a better surface finish. The bismuth added to DIN 9SMn28 steel provides an increase in machinability similar to leaded steel.

6.4.2 Effect of Non-Metallic Inclusions

According to (DINIZ et al., 2006) these inclusions are manganese sulphide and are obtained by including enough sulphur to combine with manganese and iron, forming a series of manganese and iron sulphides (mainly the former), which are insoluble in steel (this type of steel is called resulphurised). MnS inclusions can be globular or can become elongated during the rolling of the steel.

(DINIZ, 2006) found that in any case, they favour machinability because they cause the formation of a brittle chip and act as a kind of lubricant, preventing the chip from adhering to the tool and destroying its cutting edge (APC), as well as improving the surface quality of the workpiece. When machining these types of steel, the cutting speed can be greatly increased (up to double) compared to that used with non-resulphurised steels. Table 6 shows the chemical composition of some steels of this type. We can comment on the composition:

Table 6 - Chemical Composition of Some Resulphurised Steels

SAE type	% Carbon	% Manganese	% Phosphorus	% Sulphur
1110	0.08-0.13	0.30-0.60	0.040 max	0.08-0.13
1112	0.13 max	0.70-1.00	0.07-0.12	0.16-0.23
1118	0.14-0.20	1.30-1.60	0.040 max	0.08-0.13
1125	0.22-0.28	0.60-0.90	0.040 max	0.08-0.13

Source: DINIZ, 2006

1) The carbon content of 11XX steels can be as low as 0.55%, with higher carbon steels being preferred in applications where, in addition to machinability, better mechanical properties are also required. For example, 1137 steel often replaces 1045 and 1050 steels in parts such as generator shafts, certain types of screws, gears and similar parts which, in addition to machinability (parts that undergo a lot of machining), must be hard, wear-resistant and tough. 1137 steel has a machinability index of 70 (DINIZ, 2006).

2) In some low carbon types, phosphorus can be introduced in addition to the normal levels. This element dissolves in the ferrite, whose hardness and mechanical strength are increased, which improves machinability as it promotes chip breakage and makes it more difficult to form a false cutting edge. However, 0.12% P should not be exceeded, otherwise its negative effects could prevail (DINIZ, 2006).

CHAPTER 7

FINAL CONSIDERATIONS

Machinability is a set of characteristics that make up the metal's response to the stresses encountered during machining. As such, it is influenced by the specific operation in use, the maximum cutting speed for a given surface finish, the amount of material pulled out, tool wear, the metallurgical conditions of the metal, etc.

The metallurgical aspects involved in obtaining reproducible machinability properties vary according to the material to be machined. The increase in the content of chemical elements such as carbon, sulphur and lead, due to metallurgical phenomena, has a strong influence on the wear of the cutting tool, proving beneficial to improving machinability.

Carbon is the element that directly influences the machining of steel through hardness. Therefore, the greater the amount of carbon, the greater the hardness of the material, caused by the increase in cementite in the microstructure of the steel, which in turn has a high hardness and accelerates the wear of the cutting tools.

It was observed that as the cutting speed increases, there is a reduction in the machining force. Depth of cut and feed cause a greater increase in machining force than the other machining parameters during the turning operation.

The cutting temperature increases as the speed increases. Small feed rates lead to lower temperatures. The machinability of a steel is strongly influenced by the type of tool used, the choice and correct use of cutting fluid and the determination of tool position angles.

SUGGESTIONS FOR FUTURE WORK

- To study the effects of alloying elements on the machinability of steels;

- To evaluate the influence of cutting parameters on chip formation during the machining of steels;

- To study the surface integrity of steels in various machining processes;

- To analyse or study the machinability of steels at high cutting speeds.

- To study the effect of inclusions on the machinability of steels.

BIBLIOGRAPHICAL REFERENCES

AMARO, J. P. M. et al. **Development of low alloy steels for mechanical construction with improved machinability.** In: XXXVII ABM Annual Congress, Rio de Janeiro, 1982, p. 271-281.

BARRIOS, André Nozomu Sadoyama. **Thermal Modelling for Temperature Evaluation in the Milling of Mould and Die Steels.** Master's thesis submitted to Universidade Estadual Paulista. Faculty of Engineering of Ilha Solteira, 2013.

AKASAWA, T.; SAKURAI, H.; NAKAMURA, M.; TANAKA, T.; TAKANO, K. Effects of freecutting additives on the machinability of austenitic stainless steels. Journal of Materials Processing Technology. p. 143-144, 2003, 66-71.

BAPTISTA, A. L. B. **Metallurgical aspects in the evaluation of steel machinability.** Revista Escola de Minas, V. 55, n.2, p. 103-109, 2002.

CALLISTER, W. D. Fundamentals of materials science and engineering: an integrated approach. Rio de Janeiro: LTC, 2006.

CASAGRANDA, M. V. Study of the Machinability of AISI 303 Austenitic Stainless Steel. 2004. 25f. Monograph (Mechanical Engineering Course Conclusion Paper) - Department of Mechanical Engineering, Federal University of Rio Grande do Sul, Porto Alegre, 2004.

CHIAVERINI, Vicente. **Steels and cast irons.** Brazilian Association of Metallurgy and Materials, 2005.

COSTA, Dalberto Dias et al. **Analysis of turning parameters for hardened steels.** State University of Campinas, Master's Thesis, 1993.

DA SILVA, MÁRCIO AURÉLIO, et al. **Residual Force Analysis in the Machining of BNT 1045 Steel.** In: 6th Brazilian Congress of Manufacturing Engineering, Caxias do Sul-RS, 2011.

DINIZ, A. E.; CUPINI, N. L. **Study of the drilling process for austenitic stainless steels.** Metalurgia - ABM, v. 41, n. 349, p. 881-886, dez. 1986.

DINIZ, Anselmo Eduardo; MARCONDES, Francisco Carlos; COPPINI, Nivaldo Lemos. **Materials Machining Technology.** Arliber Editora, 2006.

DE SOUZA, Franco Luiz Castilho, et al. **Evaluation of the chip formation process in the turning of ABNT 1045 steel.** Department of Mechanical Engineering, Faculty of Engineering - Ilha Solteira Campus, 2006.

FERRARESI, D. **Fundamentos da Usinagem dos Materiais.** São Paulo: Blucher LTDA, 1977.

SCHNEIDER Jr. G.; CMfgE Professor Emeritus **Engineering Technology, Lawrence Technological University.** 2009.

GERDAU. Metalúrgica Gerdau S.A. Available at: <http://www.gerdau.com> Accessed on: 06 March 2016.

GONZALES, G. S., CUPINI, N. L. **Metallurgical aspects in the evaluation of the machinability of SAE 12L14 steel submitted to increasing degrees of hardening by Wire Drawing.** In: XXXXVII ABM Annual Congress, Belo Horizonte, 1992, p. 171-188.

IABR. The steel industry in Brazil. Brasília, CNI, 2012. Available at: <http://www.acobrasil.org.br/site/portugues/sustentabilidade/downloads/livro_cni.pdf> Accessed on: 26 April 2016.

KLUJSZO, Luis Augusto Colembergue; SOARES, Rodrigo Belloc. Improved machinability corfac steels at gerdau açominas s.a. Steels. Gerdau Açominas S.A. Rio grande do Sul, 2004.

MALYNOWSKYJ, A. Manufacture of liquid steel in an oxygen converter. Cubatão: COSIPA, 2000. 68p.

MACHADO, Álisson Rocha et al. **Teoria da Usinagem dos Materiais.** São Paulo: Editora Blucher, 2009.

METAL TUBE. Olinda, Pernambuco, September. 2016. Available at http://metaltubo.com.br/?page=home. Accessed on: 30 September 2016.

NASCIMENTO, F. A.; RIBEIRO, M. V. **Evaluation of the machinability of AISI 410 martensitic stainless steel in turning**. Guaratinguetá, Unesp Institutional Repository, 2008.

PANNONI, Fabio Domingos. Structural steels. Açominas, 2009.

PASSOS, L. Materials Science and Technology Workbook, Faculdades Integradas Einstein de Limeira. São Paulo: [s.n.], 2006.

PIMENTEL, Marcelo Francisco. **Influence of Chemical Composition and Microstructure on the Machinability of Easy Cut Steel with Lead Addition (SAE 12L14).** Dissertation presented for a master's degree. Paulista State University, 2006.

PIMENTEL, M. F.; PRADO, E. L. do. The machinability of easy-cutting low-carbon steels. Available at <http://www.guiadasiderurgia.com.br/novosb/component/conte nt/article/104materias20/693-acos-corte-facil-baixo-carbon>. Accessed on: 10 March 2015.

SHAW, Milton Clayton. **Metal Cutting Principies**. New York: Oxford University Press, 2005.

TRENT, Edward Moor; WRIGHT, Paul Kenneth, **Metal Cutting**. Butterworth Heinemann, 2000.

WEINGAERTHER, W. L.; SCHROETER, R. B. **Machining Processes, Manufacturing by Material Removal**. 2003.

38

yes
I want morebooks!

Buy your books fast and straightforward online - at one of world's fastest growing online book stores! Environmentally sound due to Print-on-Demand technologies.

Buy your books online at
www.morebooks.shop

Kaufen Sie Ihre Bücher schnell und unkompliziert online – auf einer der am schnellsten wachsenden Buchhandelsplattformen weltweit! Dank Print-On-Demand umwelt- und ressourcenschonend produziert.

Bücher schneller online kaufen
www.morebooks.shop

info@omniscriptum.com
www.omniscriptum.com

Printed by Books on Demand GmbH, Norderstedt / Germany